ANIMAL WORLD

THE DEADLIEST ANIMALS IN THE WORLD

by Philip Wolny

BrightPoint Press

San Diego, CA

© 2024 BrightPoint Press
an imprint of ReferencePoint Press, Inc.
Printed in the United States

For more information, contact:
BrightPoint Press
PO Box 27779
San Diego, CA 92198
www.BrightPointPress.com

LIBRARY OF CONGRESS CATALOGING-IN-PUBLICATION DATA

Names: Wolny, Philip, author.
Title: The deadliest animals in the world / by Philip Wolny.
Description: San Diego, CA: BrightPoint Press, [2024] | Series: Animal world | Includes
   bibliographical references and index. | Audience: Ages 13 | Audience: Grades 7-9
Identifiers: LCCN 2023004146 (print) | LCCN 2023004147 (eBook) | ISBN 9781678206147
   (hardcover) | ISBN 9781678206154 (eBook)
Subjects: LCSH: Dangerous animals--Juvenile literature.
Classification: LCC QL100 .W65 2024 (print) | LCC QL100 (eBook) | DDC 591.6/5--dc23/
   eng/20230313
LC record available at https://lccn.loc.gov/2023004146
LC eBook record available at https://lccn.loc.gov/2023004147

CONTENTS

AT A GLANCE

- Animals use many tools to be deadly, including size, strength, stealth, and venom.

- The Nile crocodile is one of the world's largest reptiles. It hunts prey at the water's edge in central Africa.

- Nile crocodiles kill prey by performing a death roll and dragging their prey underwater.

- The grizzly bear is a large, fierce woodland predator in North America.

- Grizzlies are known for their great strength and aggression when surprised or challenged.

- The king cobra is the largest venomous snake in the world. Its range extends through much of Asia.

- King cobras mostly keep to themselves but can be deadly to animals and humans alike. They become especially aggressive when defending their nests.

- Mosquitoes may be small, but they are the deadliest animals on Earth. They live on every continent except Antarctica.

- When mosquitoes bite humans to drink blood, they can spread diseases such as malaria. Diseases spread by mosquitoes kill more than 700,000 people each year.

POWERFUL PREDATOR

It was May 31, 2020, in Yellowstone National Park. The park is one of the few places in the lower forty-eight states where grizzly bears roam freely. That day, park visitors caught a rare sight. A young grizzly bear was hunting a bison. The bear had to be careful. The bison was a bit bigger than the grizzly. It also had sharp horns. The horns can be used to defend against predators.

Suddenly, the bison charged the bear. The grizzly used its speed to avoid the attack. It was able to circle its prey and grab it from behind. The grizzly stood on its

Bison are the largest mammals in North America. They're strong enough to kill bears in self-defense.

Grizzly bears eat up to 30 pounds (14 kg) of food every day.

hind legs and held on to the bison with its

powerful forelegs and claws.

The bison did all it could to escape. It dragged the bear into a creek. But the bison got weaker and weaker. The bear eventually won the struggle. It also won a meal.

NATURE'S DEADLIEST

The grizzly bear is one of nature's deadliest predators. Predators have developed natural weapons and defenses over millions of years of evolution. They use these adaptations to hunt and protect themselves. Many predators can be deadly to animals and humans alike. Some animals kill with their sharp claws and teeth. Others use quick-acting venoms. Still others kill unintentionally with deadly viruses. Humans must understand these animals in order to coexist safely with them. Knowledge can also help the animals. Humans who fear them will leave them alone. This helps preserve their populations and **habitats**.

People can protect wildlife by respecting animals and observing them from a safe distance.

1

NILE CROCODILES

The animals of Africa depend on rivers to get drinking water. But when animals visit the water's edge, they are cautious. They have good reason to be scared. At any moment, one of nature's deadliest predators can strike. That predator is the

Nile crocodile. It is not just a threat to other animals. It can also be deadly to humans.

A MASSIVE CROCODILE

The Nile crocodile is one of the largest reptiles in the world. Nile crocodiles have

Nile crocodiles can swim at 19 to 22 miles per hour (30–35 kmh).

an average length of 16 feet (5 m). Their average weight is 500 pounds (227 kg). But they can get much bigger. The largest Nile crocodile on record was 21.3 feet (6.5 m) long. It weighed 2,400 pounds (1,090 kg).

Male crocodiles are typically larger than female crocodiles. But females can still get as heavy as 660 pounds (300 kg). Adult females are usually no longer than 13 feet (4 m).

Nile crocodiles are shaped like huge lizards. Their hides are scaly and rough. They have short legs and long tails. Their tails, bodies, and heads are very muscular.

Nile crocodiles are devoted parents. Mothers will carry their babies in their mouths to protect them.

Crocodiles have the most powerful bite of any animal in the world.

HABITAT AND PREY

Nile crocodiles are carnivores. They are found in rivers, lakes, swamps, and other wetlands. They live and feed in fresh water. They catch most of their prey at the water's edge.

As much as 70 percent of their diet is fish. The rest of their diet is made up of reptiles, birds, and mammals. They eat snakes, ostriches, and mammals such as gazelles, wildebeests, zebras, monkeys, and hippopotamuses. The most common land animals on their menu are antelope.

NATURAL WEAPONS

Nile crocodiles have very powerful jaw muscles. Gregory M. Erickson of Florida

Nile crocodiles are cold-blooded. This means that their bodies don't regulate temperature. They sunbathe to stay warm. They sleep in caves on riverbanks when it's too hot.

State University studies crocodiles. He says they have not changed much in tens of millions of years. Prehistoric crocodiles are thought to have been at least 20 feet (6 m) long. Their bites may have been as powerful as that of the *Tyrannosaurus rex*.

Erickson explains how crocodiles' powerful bites have helped them survive. He says, "That's why I think they've been so successful. They seized on a remarkable design for producing bite force and pressure to occupy **ecological niches** on the water's edge for 85 million years."[1]

AN APEX PREDATOR

Apex is a Latin word that means top or peak. Apex predators are at the top of the food chain in their **ecosystems**. They prey on other animals, but they have no natural predators that prey on them. The Nile crocodile is the apex predator of Africa's freshwater areas.

Nile crocodiles have narrow mouths compared with alligators. Their mouths are filled with large, slightly curved teeth. Crocodile teeth are not meant for cutting and biting. They are not sharp. They are blunt. Such teeth are meant to hold their prey.

A crocodile can take down even large mammals. It can make a meal of a 600-pound (272-kg) wildebeest. First, the crocodile will bite down on the victim's neck. The bite might break the neck. It may also cause the animal to bleed to death. Then the crocodile will drag its prey into

Nile crocodiles can sleep with one eye open and watching. This keeps them safe from being ambushed.

the water. Its powerful jaws keep its victim underwater until the prey drowns.

The crocodile's jaws and teeth help it with its other fearsome weapon. This weapon is a move known as the death roll. The crocodile will start spinning or rolling as it holds on to its prey. This motion can rip large chunks off its prey's body.

Crocodiles do not chew their food. The death roll helps them break prey down. They can then swallow and digest their meal. A crocodile can do a death roll on land as well as in water. The motion also makes the prey dizzy and panicked. This makes it easier to control and kill. Crocodiles will also death roll when fighting other crocodiles for territory or a mate.

NILE CROCODILES AND HUMANS

Humans are also at risk around these fearsome animals. Crocodiles rely on the element of surprise. Evolution has given

Nile crocodiles are stealthy predators. If they stay completely still, they can hold their breath for up to two hours.

them eyes, ears, and nostrils on top of their heads. This allows them to see, hear, smell, and breathe while underwater.

Nile crocodiles can be very aggressive. They sometimes consider humans a territorial threat. Many attacks are unreported. It is estimated that Nile crocodiles kill 200 humans every year. Many more suffer serious injuries. Attacks happen when humans and crocodiles live close to one another.

One of these deaths happened in 2006. A doctor named Richard Root was taking a wildlife river tour in eastern Botswana. His wife, Rita O'Boyle, followed in a nearby boat. She says, "All of a sudden, the canoe shook. Dick was pulled over [the side] and didn't come up again. It was very bad."[2]

2
GRIZZLY BEARS

Grizzly bears are a type of brown bear. Brown bears living inland are usually known as grizzlies. Grizzlies once lived all over the western United States, Canada, and Mexico. Now they're less common. In 2021, the US Fish & Wildlife Service (USFWS) estimated that at least

1,923 grizzlies lived in the lower forty-eight states. Alaska had about 31,000 grizzlies. Canada had about 29,000.

A WOODLAND GIANT

Food can be hard to get for inland grizzlies. This makes them competitive. They have a

Baby bears are called cubs. Grizzly cubs stay with their mothers for about two and a half years.

reputation for being aggressive with both humans and other animals. This includes other bears.

Adult grizzlies are big. They weigh between 500 and 900 pounds (227–408 kg). They can reach almost 9 feet (2.7 m) in height when they're standing on their back feet.

Grizzlies get their name from their grizzled hair. Streaks of blond-tipped fur grow on their backs and shoulders. Grizzlies are known for their large heads, short ears, and short tails.

Grizzly bears don't hibernate. They instead fall into a deep sleep for about seven months. They can still wake up if startled, so it's not hibernation.

Grizzlies are also known for the hump of muscle on their backs. This gives them strength for digging. Bears dig to look for food. They also dig dens for the winter.

BUILT FOR SURVIVAL

Grizzlies have big paws. Their front claws are at least 2 inches (5 cm) long. Some grizzlies have claws as long as 4 inches (10 cm). These are twice as big as those of their black bear cousins. These claws are good for digging, hunting, and fighting.

The bite of a grizzly bear is stronger than that of a lion. It is powerful enough to crush a bowling ball. Grizzly teeth can handle many different food sources. They can crush plants and tear flesh.

Some researchers estimate a grizzly has the strength of up to five humans. In 2006,

Grizzly bears are deadly on land and in water. They are expert swimmers. Their oily fur and high levels of body fat help them float in the water.

National Geographic hired engineering students to test bear strength. One test involved a 700-pound (320-kg) metal dumpster. The students' adviser, Professor Doug Cairns, said, "It was like a beach ball to them. They could roll it over and over. It took a minimum of two people a concerted effort to tip it."[3]

HUNTING AND SCAVENGING

Grizzly paws are also strong. Gary Brown, author of *The Bear Almanac*, writes, "No animal of equal size is as powerful. A bear may kill a moose, elk, or a deer by a single blow to the neck with a powerful foreleg, then lift the **carcass** in its mouth and carry it for great distances."[4]

Grizzlies are opportunists. This means they eat what is available. A grizzly will often pick easier targets over tougher ones. A hungry bear may try to hunt and kill any animal, even very big ones. But a typical grizzly often picks the young of larger animals. Older or wounded animals are also easier prey. Grizzlies may sometimes **scavenge** the kills of wolves and other predators. They also feed on animals that die during the winter.

SUPER SMELL

The grizzly's sense of smell is powerful. Its nose is about seven times as strong as a bloodhound's and 2,100 times as strong as a human's. Grizzlies can smell things as far as 2 miles (3 km) away. This helps them detect food, prey animals, and danger.

Salmon is an important part of a grizzly bear's diet. In late summer, grizzlies can eat more than 100,000 calories of salmon in a single day.

A grizzly's teeth, claws, and strength are also important for the plant part of its diet. Grizzlies dig up root vegetables and mushrooms. They eat honey, nuts, and fruits. They also eat grasses, flowers, and other plants. They use their strength to

move fallen trees and other obstacles to

get to food. Insects also make up a big

part of a grizzly's diet. They can eat

thousands of moths and ants in a few

hours. Grizzlies also scavenge the hidden

food stores of other animals.

GRIZZLY VS. HUMANS

The USFWS listed grizzlies as an

endangered species in 1975. Since then,

their populations have slowly recovered. But

having more bears also means there have

been more human encounters with bears.

In late September 2019, four human hunters were injured by grizzlies in Montana in three separate attacks. All of the hunters survived. They were lucky. There are few good choices in such a run-in. Grizzlies can run up to 35 miles per hour (56 kmh).

There have been only about 180 fatal bear attacks in North America since the late 1700s. Attacks are more likely in autumn. Bears must store as many calories as they can before winter. Surprising a hungry grizzly is extremely dangerous.

The more human contact wild bears have, the more likely the bears are to attack.

Some national parks have bear-proof storage lockers to keep food from attracting bears. Staff have to change the design whenever bears figure out how to open the lockers.

Leaving food outside is dangerous. This attracts more bears. Humans must secure their food carefully to avoid attracting bears. Wildlife officials work to educate the public about proper safety around bears. This prevents tragic attacks from happening.

3

KING COBRAS

The king cobra is the world's largest venomous snake. The name *king cobra* comes from this animal's ability to hunt and eat other cobras.

Most adult king cobras are between 9 and 12 feet (2.7–3.7 m) long. One measured in Malaysia in 1937 was more than 18 feet

(5.5 m) long. Adults are often yellow, green, brown, or black. They often have white or yellow bands. These bands are also known as crossbars or chevrons. The bands appear on the snake's back. Sometimes a snake will also have bands on its underside.

King cobras can't hear the way that humans can. Their ears instead sense ground vibrations. This helps them tell if something is coming toward them.

An average adult king cobra weighs about 13 pounds (5.9 kg).

King cobras know how to appear dangerous. They make themselves look and sound intimidating when they feel scared. This is called a **threat display**. It warns animals and people to stay away. The king cobra has several parts to its threat display. One part is its hood. The king cobra has flaps of skin on its neck. The snake can expand its ribs and muscles at the sides of its neck to form a flared hood. A king cobra can also raise as much as one-third of its body off the ground. This standing motion

The hole in a king cobra's mouth is called the glottis. It lets the cobra breathe while it's eating. It's also what makes the snake's hissing sound.

lifts the snake's head straight into the air. It can make the snake taller than the average human. It also lets the snake look over bushes and tall grasses.

Sound is also part of the king cobra's threat display. A king cobra can puff and hiss as a warning. The hood and lifting motion make the snake look bigger. Its low-pitched hissing sounds like a growl. Together, this threat display scares away predators. It even scares away other snakes.

FEEDING TIME

The king cobra's hollow fangs are almost half an inch (12 mm) long. They deliver a deadly venom. The venom works quickly. It paralyzes the snake's prey. The poison

King cobras may go months without eating. It takes a long time for them to digest their food.

makes the snake's victim unable to breathe or move. Toxins in the venom start to break down the prey's body. The king cobra then swallows and digests its prey.

The king cobra mainly eats other snakes. Its diet includes Asian rat snakes, pythons, Indian cobras, and even smaller king cobras. It mostly favors snakes that are 10 feet (3 m) long or shorter. Some king cobras eat only a single species of snake.

WRESTLING, KING COBRA–STYLE

Stakes are high during the king cobra breeding season. Male cobras compete for mates and territory. Aggressive males challenge each other to a form of wrestling. The snakes curl around each other. They try to push each other's heads to the ground. The one who pushes the other's head down is the winner.

VICIOUS AND VENOMOUS

King cobras are found in southern and southeastern Asia. They live in forests, bamboo thickets, farmland, and mangrove swamps. They normally stay near streams. King cobras spend about a quarter of their lives in trees or bushes.

The king cobra is the only snake that builds nests for its eggs. A mother king cobra builds a nest of leaves and soil. She lays up to fifty eggs. She covers the eggs with more leaves. She then lies on top of them to incubate her young. This incubation period can last two months. The father

King cobras smell using their tongues. Their tongues are forked to help them determine where a smell is coming from.

usually stays close by. Experts believe king cobras pair for life with a single mate.

It is during this period of caring for their eggs that king cobras are most dangerous. Females are aggressive toward any humans who approach their nest or young. Their powerful eyesight can see a person or other threat from nearly 330 feet (100 m) away.

King cobras do not attack humans very often. They kill five or fewer people throughout their wide habitat per year. But humans must always be careful in this snake's habitat. King cobras can attack if they are cornered or surprised. The king cobra can deliver enough venom in a single attack to kill an elephant. This is enough to kill about twenty humans.

A person bitten by a king cobra must seek immediate care. They must get to a hospital or clinic as soon as possible. An untreated bite can be fatal in 50 to 60 percent of cases. Multiple bites are even more dangerous. Victims need quick access to **antivenom** and other treatments.

FROM POISON TO PRESCRIPTION

A king cobra's venom can kill people, but it can also be used in medicine. Scientists use the snake's venom to make powerful painkillers.

King cobra venom can cause intense pain and blurry vision. Bite victims may feel drowsy and dizzy. The victim can become fully paralyzed. The person might go into a coma before their heart and lungs fail.

Snake expert George Van Horn owns a reptile zoo in St. Cloud, Florida. The zoo uses snakes to make antivenom. Van Horn was attacked by one of his snakes. He says, "That snake was my best venom producer, and he bit me four times before I got him off."[5] The attack left him with only partial use of his arm. He spent thirty-three days in the hospital.

4

MOSQUITOES

Some creatures are known for their venom or size. Others are feared for their strength and claws. But the deadliest animal on Earth is not the biggest or the strongest. It is one of the smallest: the mosquito.

FEEDING MACHINES

Mosquitoes are found in almost every part of the world. They live on every continent but Antarctica. There are more than 3,500 types of mosquitoes worldwide. Adult mosquitoes live about two to four weeks. Their life span depends on how hot and

Mosquitoes have been around for at least 400 million years.

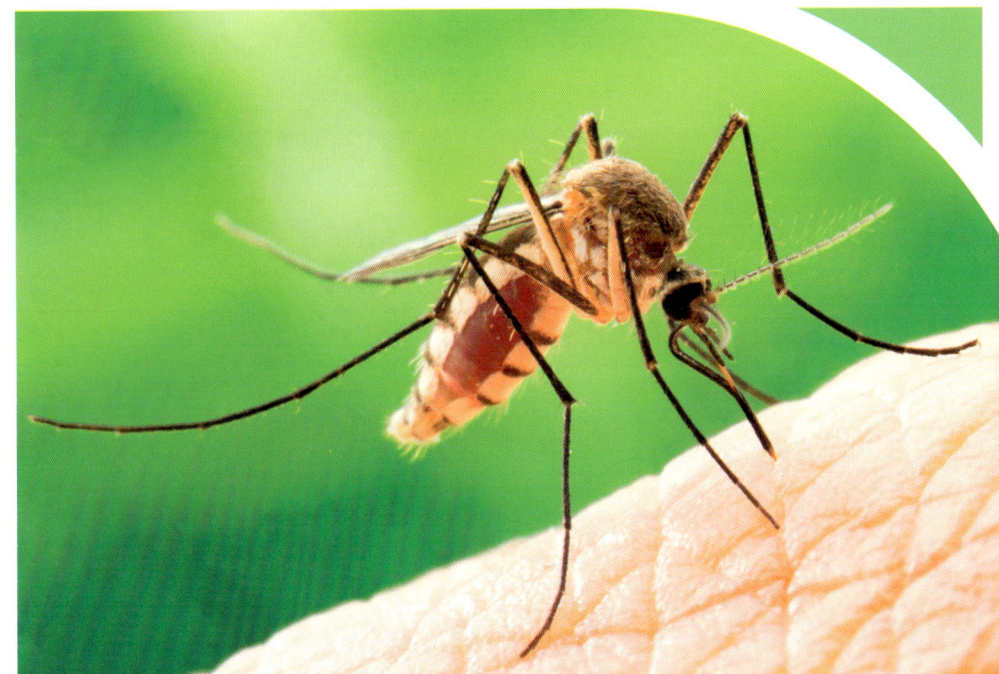

Mosquito eggs hatch into larvae. These larvae are called wrigglers. They live in water.

humid their environment is. It also depends

on the species of mosquito.

Mosquitoes' long antennae help them

find their prey. These organs follow

air movements to track humans and

animals. They can detect carbon dioxide

from breath. Mosquitoes can feel the gas from more than 100 feet (30 m) away. Next to the antennae are the mosquitoes' palps. These organs can sense odors. Mosquitoes can detect nearby humans by smell. Their large eyes also help them find food sources. Biologist Omar Akbari says, "They're really the ultimate predator. You can't find a single person on Earth that hasn't been bitten at least once."[6]

Mosquitoes also have tools that help them eat. They have a long mouth part called a proboscis. This long, sharp organ breaks through a human's skin.

The proboscis acts like a straw. Blood flows from the victim to the mosquito.

BLOOD MEALS

Only female mosquitoes drink blood. Both males and females feed on fruit juices and flower nectar. But females need blood to produce their eggs. Only about 6 percent of mosquito species draw blood from people. Of these, about half do not carry diseases. These are called nuisance mosquitoes. Their bites cause only mild itching and irritation. This reaction comes from a chemical the mosquito makes.

In places where mosquitoes spread deadly diseases, some people put mosquito nets over their beds. These nets keep out mosquitoes and can be coated in insecticides.

This chemical thins blood and stops it from clotting. This makes it easier for the mosquito to suck up the blood.

Sometimes mosquitoes drink blood from someone with a virus. That virus can sometimes pass to the mosquito. The mosquito can spread only germs that can reproduce inside the mosquito's body. It takes about two to three weeks for the germs to multiply. They move from the mosquito's body to its salivary glands.

SCIENCE TACKLES THE MOSQUITO PROBLEM

Scientists have tried many experiments to prevent mosquitoes from spreading illnesses. They have tried to change mosquito genes to prevent breeding. They have also tried to disrupt the insect's sense of smell. But mosquitoes have shown an incredible ability to adapt.

When this happens, the saliva in the mosquito bite can pass germs to humans.

A TINY TERROR

About a thousand people a year die from crocodile attacks. More than 50,000 die worldwide from snake venom. But no creature causes more human disease and death than the mosquito. According to the World Health Organization, mosquitoes are responsible for about 725,000 deaths a year. Many more people get seriously sick. Survivors can have permanent health problems.

DEADLIEST ANIMALS IN THE WORLD

The number of people each group of animals kills per year is represented by the area of the circle.

Mosquitoes[1]
725,000

Humans[2]
475,000

Crocodiles[2]   Hippos[2]   Lions[2]   Sharks[2]

1,000   500   250   5

Snakes[2]
50,000

Dogs[2]
25,000

1. "Mosquito as Deadly Menace," Pfizer, October 4, 2016. www.pfizer.com.

2. Jessica Learish, "The 24 Deadliest Animals on Earth, Ranked," CNET, October 15, 2016. www.cnet.com.

Though smaller than a fingernail, the mosquito is by far the deadliest animal in the world.

It is not the mosquitoes themselves that kill. It is the many different diseases they spread. The biggest killer by far is malaria. Malaria kills about 600,000 people yearly. Other deadly mosquito-borne diseases include yellow fever, dengue, encephalitis, West Nile virus, and Zika virus.

Some animals use their claws to defend themselves. Others use their venom. Some animals are huge. Some you could squish between your fingers. But all animals have found ways to survive. People can stay safe by learning more about animals and giving them space in the wild.

GLOSSARY

antivenom

a substance used to treat poisoning from snake venom, often made using snake venom

carcass

a dead body of an animal

ecological niches

the roles and places of species within an ecosystem

ecosystems

complex systems made up of the plants and animals in an area and their environment

habitats

the places where particular animals or plants naturally live and grow

scavenge

to look for food or other resources

threat display

any action or set of actions animals perform to scare away rivals or predators

SOURCE NOTES

CHAPTER ONE: NILE CROCODILES

1. Quoted in Brian Handwerk, "Crocodiles Have Strongest Bite Ever Measured, Hands-On Tests Show," *National Geographic*, March 15, 2012. www.nationalgeographic.com.

2. Quoted in Carol M. Ostrom, "Botswana Trip Turns Fatal for Disease Expert from Seattle," *Seattle Times*, March 22, 2006. www.seattletimes.com.

CHAPTER TWO: GRIZZLY BEARS

3. Quoted in Tracy Ellig, "MSU Researcher Tests Grizzly Bear Strength for National Geographic Documentary," *Montana State University*, July 3, 2006. www.montana.edu.

4. Quoted in Gary Brown, *The Bear Almanac: A Comprehensive Guide to the Bears of the World*. 2nd ed., Guilford, CT: Lyons Press, 2009.

CHAPTER THREE: KING COBRAS

5. Cheri Henderson, "Story of a Herpetologist," *Orlando Magazine*, October 12, 2020. www.orlandomagazine.com.

CHAPTER FOUR: MOSQUITOES

6. Quoted in Katherine J. Wu, "You Have No Idea How Good Mosquitoes Are at Smelling Us," *Atlantic*, August 18, 2022. www.theatlantic.com.

FOR FURTHER RESEARCH

BOOKS

Sue Gagliardi, *Endangered Sharks*. San Diego, CA: BrightPoint
 Press, 2023.

Anita Ganeri, *Deadly Animals in the Water*. New York: PowerKids
 Press, 2022.

Karen McGhee, *World's Most Deadly Animals*. New York: Gareth Stevens
 Publishing, 2022.

INTERNET SOURCES

Priyanka Roy, "Steve Backshall on Filming the World's Most Dangerous
 Animals," *Telegraph Online*, January 20, 2021.
 www.telegraphindia.com.

"What Are the World's Deadliest Animals?," *BBC*, June 15, 2016.
 www.bbc.com.

Leoma Williams, "10 Deadliest Animals to Humans," *Discover Wildlife*,
 June 25, 2022. www.discoverwildlife.com

WEBSITES

BBC Wildlife
www.discoverwildlife.com

Founded as *Animals Magazine* in 1963, *BBC Wildlife* publishes articles on wild animals, plant life, and the environment.

National Geographic Kids
https://kids.nationalgeographic.com

The National Geographic Society is a nonprofit and educational organization dedicated to spreading awareness and knowledge about geography, world cultures, natural science, and related topics. It has published the *National Geographic Kids* magazine as a resource for young readers since 1975.

World Wildlife Fund (WWF)
www.worldwildlife.org

The World Wildlife Fund is one of the world's leading nature conservation nonprofits. Active in almost 100 nations, it develops solutions to preserve ecosystems and educates the public, local communities, and governments about nature and conservation.

INDEX

IMAGE CREDITS

ABOUT THE AUTHOR

Philip Wolny is a writer, editor, copyeditor, and animal lover. He lives in Florida with his wife and daughter. The unique flora and fauna have proved the most fascinating and appealing features of the Sunshine State, with most of Florida's remarkable biodiversity thankfully far from dangerous or deadly.